WAR AFTER THE MILITARY

TAMMIE OTUKWU

www.TrueVinePublishing.org

War After the Military
Tammie L. Otukwu

Published by True Vine Publishing Company
P.O. Box 22448, Nashville, TN. 37202
www.TrueVinePublishing.org

Copyright © 2020 by Tammie L. Otukwu
ISBN—13: 978-1-7336315-4-9

All rights reserved. No part of this book may be reproduced in any form or by any electronic or mechanical means without permission in writing from the publisher, except by a reviewer who may quote brief passages in a review.

Printed in the United States of America—First Printing

To place orders for more books or to set up speaking engagements, contact the author at www.TammieOtukwu.com

Endorsement

Nothing appears as it seems. However, when thought transfers from the head of man to the heart, it's Divine. Divine intervention is paramount when translating interpretation. The interpretation has the power that only God can render, then divine order is the ORDER!

Devoted Wife, Mother, Author and over 20 years of service retiree SFC Tammie Otukwu has penned an excellent work to express her divine view on "War After the Military", which speaks to "it is what it is but it ain't what you think"...

#IJN
~Calvin V. Jennings
"Popz"

DEDICATION

I dedicate this book to my Superheroes: my Mother, Mary Nell and my Aunt Dorothy. Each one taught me about faith, love, family and perseverance. Thank you both for your prayers, belief in my dreams, support and strength. I am eternally grateful.

FOREWORD
J. D. PAIR

As a fellow veteran with over 30 years of service, I was a bit intimidated when asked to write the forward for this glorious book. Upon much thought and soul searching, I agreed as a "Battle Buddy" (who civilians refer to as a friend), and a fellow professional who works with other fellow veterans. It was my honor. Given the supreme achievements of the author within the military and after, it was apparent to me that she had a story which had to be told.

This story gives a keen insight into what happens from a mental and physical point of view that transitioning military personnel experience often. Everyone thinks or expects a military person to always present themselves as "dress right, dress" or all together, but the truth is much deeper and more complex.

The public at large hasn't yet begun to fully understand that "transitioning" out of the military doesn't mean giving up being a Soldier, but rather learning to pilot our warrior responses depending on the circumstances. The book gave me insight into understanding myself and the patients I work with as a mental health professional. There are many books on how to navigate being in the military; few gave insight or rarely delved outside of a professional setting. The author cuts to the chase and cleverly wades through all the misconceptions we thought we knew about being a Soldier and a civilian.

The word "war" can mean many things such as: a war of words, war of the mind, or war of the body. However, the author gives her unabashed take on what the word "war" meant to her; meaning a transition, an evolution, growth, and a process out of a life and experiences few can truly understand. I feel when you read this book you will get a firsthand account and true understanding of a "War after the Military".

Prelude

The purpose of this book is to honor the many military service members who proudly served our country. I also honor those who are currently serving in the armed forces. It is my hope that this work will address questions about the military life of a Soldier as well as provide valuable resources for Veterans and their families. I speak from my own perspective as an African-American woman about my life, military journey, and the years of adjustment after the service. I hope that my writing will be a blessing to many as it has been a healing for myself. This book may not fit everyone's situation. Everyone's path is unique. My journey has not always been easy.

I have spent over 26 years in the Army, traveling within the United States and overseas, working with and meeting different types of people. I have worked with homeless Veterans, church members, family members and Veterans in crisis. I have a Master of Psychology which taught me a basic understanding of human nature. I also have a Master of Management degree which taught me how to work with groups of people.

I wish that I could say that my time spent in the military, life experiences, or education would have prepared me for leaving the service, but it did not. Nothing but God prepared me for the feelings of isolation, the depression, sleepless nights, agitation, lack of trust, and many other emotions I experienced during that time.

Creed of the Noncommissioned Officer

No one is more professional than I. I am a Noncommissioned Officer, a leader of soldiers. As a Noncommissioned Officer, I realize that I am a member of a time honored corps, which is *known as "The Backbone of the Army."*

I am proud of the Corps of Noncommissioned Officers and will at all times conduct myself so as to bring credit upon the Corps, the Military Service and my country regardless of the situation in which I find myself. I will not use my grade or position to attain pleasure, profit, or personal safety.

Competence is my watch-word. My two basic responsibilities will always be uppermost in my mind--accomplishment of my mission and the welfare of my soldiers. I will strive to remain tactically and technically proficient. I am aware of my role as a Noncommissioned Officer. I will fulfill my responsibilities inherent in that role. All soldiers are entitled to outstanding leadership; I will provide that leadership. I know my soldiers and I will always place their needs above my own. I will communicate consistently with my soldiers and never leave them uninformed. I will be fair and impartial when recommending both rewards and punishment.

Officers of my unit will have maximum time to accomplish their duties; they will not have to accomplish mine. I will earn their respect and confidence as well as that of my soldiers. I will be loyal to those with whom I serve; seniors, peers and subordinates alike. I will exercise initiative by taking appropriate action in the absence of orders. I will not compromise my integrity, nor my moral courage. I will not forget, nor will I allow my comrades to forget that we are professionals, Noncommissioned Officers, leaders!

CHAPTER 1
SHAPING THE DREAM

There were three people who shaped my life: My father, my mother, and my grandmother. They were all completely different in their approach to raising me, all showed love in different ways, and all showed unique approaches to life that I can finally look back upon as the seeds to my success. From them I learned the power of perseverance and focused determination in the face of overwhelming odds; finding patience and strength in the face of despair, and faith and determination to achieve any goal. Had I seen one without the other two, I would not be who I am today.

I was the youngest and the most stubborn of my parents' children. As a little girl with two sisters and one brother, I had to hold my own. We were a tough but loving family and my father, who was a Vietnam Veteran, demanded discipline. What he said was law. He didn't cut us any slack just because we were children and definitely not because we were girls. Our clothes had to be pressed, shoes polished, and by God when we made our beds, he wanted to be able to bounce that quarter. Hooah!

As a child, I hated how strict my father could be. However, I can understand why he was as driven and tough as he was. My father was the oldest of ten siblings, so he was always responsible for taking care of others. Life as a black child in the 1930s was not easy, so he didn't have time to play games and enjoy the leisure our kids have today. My father dropped out of school at the age of 12 to help earn money for his family. Instead of being a kid, he was up in the early hours of the morning working until the late hours of night.

In 1954, at the age of 26, my father decided the best way to support himself was to enlist in the Armed Forces. He committed himself to becoming a career Soldier. My father was always goal oriented. In spite of his lack of formal education, he was a very smart man. He was a strategic thinker and had laid out a plan for his future. However, in 1965, the US officially entered the Vietnam War, and my father was shipped off to fight. I was just a toddler when my father was deployed, so I don't remember him leaving. But I do know he didn't come back the same.

After returning home, my father would line us up for family time to instill his doctrine of life. He would give us long lectures about the importance of education and hard work. He would also warn us about how untrustworthy people were. A lot of his conversations included being very strategic about life and "watching out for people". Perhaps he wanted to protect his daughters by scaring us and making us distrust strangers, or maybe the horrors of war made him lose his trust in humanity.

My father worked hard to progress in the military regardless of the many racial barriers he would face. I remember him coming home one day in a rage. I listened from my room as he blew off steam about the discrimination he was facing.

"I can't believe they have passed me up again!" he fussed.

"What's wrong baby?" my mom tried to console him.

"My commander refused my promotion again. I work twice as hard as these white boys and they've been promoted twice as much as I've been."

As I listened to the burning anger in his voice, I asked myself, *"why does he keep working for them. Why doesn't he just quit and find another job?"* Much to my own confusion, I witnessed him get up the next day and put on his uniform with pride and walk out to a job that showed him no respect. When I joined the Army, I got a chance to witness first-hand the challenges my father confronted. I realize now that I was being groomed for my future through my father's example.

I was blessed to see an example of a man ramming his mind, body, and soul into the brick wall of racial discrimination and resistance in order to clear the path for others like his own daughter to achieve greater careers in the Army. I consider my father a trailblazer. He never stopped fighting for his dream and eventually retired in 1974 at Ft. Hood, TX as an Army Sergeant First Class E-7.

***

My mother was a homemaker which was common in that time. Just like my father, my mother was also the oldest of 10 siblings. Mom was a caretaker and saw to the wellbeing of everyone in her family. She was a doting mother who loved her family with all of her heart. She would wake up early in the mornings before anyone in the family and start preparing for the day. My mouth still waters as I think about the aroma of bacon, eggs and grits floating through the air, bidding us to arise. Breakfast would be hot and steaming on the table when she would call us.

"Kids! Time to wake up. Come eat."

As my siblings and I would come stumbling out of the bed, we would be greeted by mom in the living room watching the morning news and starching my father's military green uniform with great care. The creases in his shirt were sharp enough to slice bread. My mother took great pride in caring for her family—just as much as my father took pride in his service in the Army. My mother understood that we needed her love, patience, and nurturing just as much as we needed our father's discipline. This symmetry of nurture and discipline provided the stability that I needed as a child.

Being a spouse of an enlisted person takes a tremendous amount of strength and patience. When my father was called on his many deployments or worked the long hours, my mother was left to take care of everything by herself. She was the cook, the cleaner, the protector, the disciplinarian, nurturer, the banker and much more. She was everything we needed all wrapped up in one person. My mom kept the family together.

Strangely, Mom never learned how to drive. She would walk miles for errands such as grocery shopping.

"Momma, why won't you learn how to drive so that we can just drive to the store?" we would ask.

She would laugh and say, "Why would I want to do that. I wouldn't get a chance to spend this time with you."

What I couldn't comprehend at that young age was that my mom cherished us so much because she knew what loss meant. She had lost two children at very young ages. I realize now, my mother enjoyed those walks with us because it was our time to slow down and be together; a time for her to instill the knowledge and values she felt were important for her children, just as my father would lecture me and my siblings. She would teach us about having faith through life lessons. I didn't know it, but she was preparing me for life as a wife and mother.

She was teaching me patience through those long walks. She exhibited strength to keep moving forward in life after the most heartbreaking and devastating experiences in a mother's life. My mother exhibited the strength of a warrior through her doting, nurturing, and loving smiles and hugs. I would use her example to push me through the Army and through life as a wife and mother.

***

When I became a teenager my mother and I had to move to Georgia to live with my grandmother. (I will talk more about this in Chapter 2). Life in the south was very different from living in Texas on the military base. The world was slower. The hustle and bustle of uniformed Soldiers was replaced with calm country roads and gravel driveways. No one was talking military language such as: "zero 100 hours", "deployments", or "commissaries".

Not only was this cultural transition hard enough, but living with my grandmother brought its own set of challenges. My grandmother was a faith warrior, a true drill sergeant in the Army of the Lord. She was a deacon's wife, president of the usher board, head of the youth group, cook in the kitchen, and assistant to the choir. She did all of this while raising ten children. She was a master at multitasking.

Grandma would keep us in church all day on Sunday and for most of the week. We had Sunday morning service, Monday night prayer, Tuesday night choir practice, Wednesday night Bible study, the list goes on. On Sunday mornings, Grandma would get up early to cook breakfast and Sunday dinner before getting us to Sunday school. I'm not talking about some flimsy microwave dinner either. I'm talking about Fried chicken, pork chops, turnip greens, mashed potatoes, hot water cornbread, and corn on the cob! We ate like royalty.

My grandmother loved God and always taught her family about the power of prayer. Whenever it stormed outside, she would tell us to get on our knees, pray and have faith. I think that she was preparing us not just for literal rainstorms, but she knew that when we grew up and faced the metaphorical storms of life, we would hear her voice saying,

"Baby, whenever it's storming, remember to get on your knees to pray. Have faith that God is doing His work."

My grandmother taught me to be thankful for each day, not taking for granted the fact that I have awaken every morning. She would always remind us that there was someone who didn't wake up, so we should be grateful for every opportunity we have to face another day. It was this faith and positive outlook on life that shaped my belief in myself and the potential that God has given me. It was my grandmother who helped me to understand that the struggles and "storms" that I would face in life were not times to fear. Instead, they were times to thank God for "doing His work."

My family and the faith they taught me to have in God are what carried me through as a child and while serving in the military. God prepared me for those hard times by allowing me to be raised in a home with discipline, love, and great faith. When I had to train alongside individuals who were stronger, faster, and often appeared to be better than me, I remembered my father's non-stop battle to achieve his promotions in the face of discrimination. I remember that even though he was at a disadvantage, he never stopped fighting for the success he knew he deserved.

When I lost family and friends, I looked back on the strength, grace, and love of my mother who lost two young children. I remembered how she pressed through the pain and heartache because she knew she had other children who were depending on her. When I doubted myself and when others doubted me and even when my faith waivered, I remembered my grandmother was praying for my strength and I heard her voice.

"It's storming baby. Get on your knees to pray. Have faith that God is doing His work."

My foundation of faith was often the only thing that I had to stand on when times were uncertain. My journey through the military was always possible through the gift of these three wonderful people.

Chapter 2
Homelessness

I was around nine years old when I had my first brush with homelessness. My dad was driving me and my brother to the store, and I was sitting in the back of the car gazing into the distance. As we were pulling up to a stop light, I saw a man standing on the side of the road in his army fatigues. I was excited to see who the man was. Living on an Army base, it was not unusual to see Soldiers walking the streets. Maybe I knew this man. What was this Soldier doing standing on the street? I wondered.

As we drew closer, I could tell something was off with this Soldier. His uniform was dirty and disheveled. I knew, even at that young age, that he was not representing the meticulous and detail oriented ideal of a Soldier that I had witnessed from my father and neighbors. Everything about how a Soldier dresses in uniform is detailed and regulated, from the way his pants legs are tucked into his boots to how low they hang above the shoestrings. This man's clothes were dirty and tattered. He was wearing old military clothes as if he was still in the service. It was as if he was caught up in a time that had passed him by.

I looked at my father who was a ranking non-commissioned officer. I'd seen him rake Soldiers across the coals for much less. I thought, *ooh! Daddy is going to let him have it!*

As we pulled to the light and stopped, my father rolled down his window. My shoulders tensed as I prepared to cover my ears for my father's tirade. Instead, my father pulled out a dollar and gave it to the man.

"Thank you, Sir," the man said to my father. I can remember so vividly the helpless and defeated look in his eyes.

My father nodded and said, "Carry on Soldier," affording the man the dignity a veteran should have.

The man picked up his cardboard sign and continued down the row of cars asking for money.

"Daddy, why did you give that man some money?" I asked.

"Because he's homeless baby. He doesn't have anywhere to live. He served his country, so we need to help him, now."

I remember feeling so confused about seeing this person in that condition. How could it be that a person who had served their country was now alone on the side of the road begging for money from strangers? My father had served in the military. My neighbors and friends all came from the same background as this man. Could we all become homeless like him?

My Dad always taught us that education was the key to having a prosperous life, so I knew that the military had excellent education benefits. Why didn't that man's education protect him from homelessness? All that I knew about a secure, stable life was challenged. The thought that someone could end up like this was very disturbing and puzzling to me. I remember the feeling of wanting to help this poor man, to give him food, money, or shelter.

As we drove away, I could not forget my feelings during that moment. I never knew how much of an impact that seeing this man would have on my life. I had so many questions, not knowing I would get a true understanding of this person's plight firsthand.

***

"Wake up baby. We have to move. Start getting your clothes packed." These were the words that started a new chapter in my life.

My mother woke me up early in the morning with tears in her eyes telling me that we were moving. I was accustomed to being deployed to different Army bases, but momma seemed to be taking this move pretty hard.

"Why are you crying momma?" I said as I packed my clothes.

"Just hurry baby. Keep packing."

I was somewhat excited. What new state or country would we be moving to now? I wondered.

The phone rang, and momma answered. I could hear her muttered conversation through the wall, and then momma exploded.

"No. I'm taking Tammie and we're going to my momma's house."

Going to grandma's? What is she talking about?

"We can't afford to live here! All the money is gone?" Momma screamed into the phone. "How am I supposed to feed the kids? You've gambled it all away!" she sobbed and slammed down the phone.

I didn't want to move to my grandmother's house away from my friends.

"Momma, why are we going to grandma's?" I asked in a panic.

"Baby just get your stuff. We can't stay here any longer."

"Is daddy coming?"

"No. Your daddy is going to stay here."

A yellow taxi pulled up to the front. Momma grabbed my bags and rushed them to the cab while I sat in my room crying. She came rushing back in and grabbed me by the shoulders.

"Baby, we have to go now. We can't stay here. Now get up."

I could tell by the look in momma's eyes that she meant business. I sucked up my emotions and downgraded my temper tantrum into a sullen pout as we drove off. As my house shrank into the distance, I remember the feeling of panic and fear of the unknown future awaiting me. What was life going to be like without my family being together?

Where were we going to sleep in grandma's house? I felt like my entire world was turning upside down. I kept hearing my mom's panic-stricken voice in my head saying, "The money is gone!" As she continued to stare into the distance, momma muttered under her breath, "I can't believe that man. Here I am, homeless with a child."

The word "homeless" shook me to the core. My mind immediately flashed back to that Soldier in the ragged army fatigues walking the street asking for change. I saw his sad helpless eyes, matted, unkept hair, and smelled his musty body, all over again just as clearly as the day he was standing in front of my father. Were we going to be like him?

My sense of stability was utterly destroyed. How could this be? Daddy said that if we got a good education and worked hard, we would have a successful life. So how is it that we don't have any money and we were losing our home? How is this happening?

I felt like my life was being stolen from me. My daddy, my brother, my room, my comfortable life—all snatched away in the blink of an eye.

CHAPTER 3
GROWING UP!

Grandma and my host of uncles, aunts, and cousins—a few of them who also lived with grandma—greeted us with open arms when we arrived in Valdosta, Georgia. The culture shock hit me immediately as grandma showed me and my mother to the rooms where we would be sleeping. I was used to having my own room, my own space, my own closet. Now I would be sharing a room.

Life for me would never be the same again. When you're in a military family, the checks come like clockwork. Everyone knows when payday is. But we had left Texas. We were no longer a "military family." Where would the checks be coming from now? I would soon learn that if I wanted something, I would have to work for it. Long gone were the days of getting everything my heart desired. Everyone under my grandmother's roof had to pitch in. If you weren't working, you had better be cooking or cleaning around the house.

As I started to get used to living in the south and making friends, I also was a teenager who wanted to meet and date boys. I was a freshman in high school when I first saw him. He was the cutest boy I had ever laid eyes on. He was light-skinned with an afro, tall and slender. He glanced my way but didn't notice me.

I saw him walking the hallway.

"Who is that?" I asked one of my friends.

My friend told me his name was Pedro and he was a senior at the high school and an art student. I was smitten. When the bell rang at the end of the day, I would rush out into the hallway searching for Pedro so that we could walk out of the school side by side, as if we were boyfriend and girlfriend. We never spoke. In fact, he didn't even know he was walking beside his "girlfriend". I never said a word to him directly. I never introduced myself, and I don't think he ever even saw my face for the first year of school. I would just walk beside him and look back at my girlfriends with a devious grin.

The school year ended, and he still didn't know me from a can of paint, even though we walked out of the school together almost every day. But that would change. The summer break came, and I got a job for a summer camp program. As the workers arrived, we waited outside for the supervisor to let us know our jobs, and who came walking out? It was Pedro. Tall and handsome as ever. I wasn't going to let the opportunity pass me by again. I made sure he would know my name and face that summer.

We got closer and closer as the summer went on. I learned so much about him while we worked together that summer. He attended church on a regular basis, always worked hard, had the same type of drive and ambition that my father did, and he even had plans to join the military. I wanted to spend every moment I could with him and he with me. We were so in love. I will never forget the day that he got the courage to ask my grandmother if he could take me to the movies, then the homecoming dance, and the prom. When we weren't spending time together, I missed him so much, I felt sick. *Is this what it means to be love sick?* I wondered. Turns out, what I was experiencing was not love sickness. The real name was "morning sickness."

***

Before I knew it, my life was turned upside down. I went from being a young high school student to a fiancé and future mom.

Within a year, I was married and the mother of the most beautiful baby girl in the world. Pedro, my new husband, had finished basic training and was being shipped to Germany. Luckily, I was no stranger to such deployments, so I was willing and ready to pack up so we could start our lives abroad. By the age of 18, I was raising a daughter in a foreign land.

We lived in Germany for a year. Pedro and I were young and infatuated with each other, but far too young to understand what it took to make a marriage work in a foreign land with no family support. It proved to be a huge strain on the relationship which inevitable led to its end.

I was 21 when we decided to go our separate ways. I didn't know exactly what I wanted in life, but I knew what I didn't want. I knew I did not want to go back to Valdosta, GA a single mother working at the local Winn-

Dixie for the rest of my life. I knew I didn't want my daughter to struggle.

I knew what it was like to have plenty and live a comfortable life, and I wanted my daughter to have that life too. When I thought about the wonderful days of my youth, my mind went back to living on the Army base. Pedro was moving up in the Army ranks and was starting to make a good living. It was obvious to me what I needed to do.

Chapter 4
The Enlistment

I began my military journey at the age of 22 while living in Hinesville, Georgia. I needed a change in my life and wanted to challenge myself. Raising my daughter in Germany was my greatest pride and joy, so the decision to enlist was not as easy as I thought. I would have to be away from my daughter for two months.

I spoke with my mother about my decision to enlist. I'll never forget what she said,

"Make sure you find a church home. Most importantly, trust in the Lord."

Her words encouraged me to trust that God was going to take care of me and my daughter while we were apart. My mother survived the loss of two children, so I knew I could survive a couple of months.

I met with my recruiter and signed my initial paperwork. I had a few months before I really had to commit, so I still had time to change my mind. I procrastinated on my decision as long as I could, but I knew this would be the best way to provide my daughter with the life she deserved. Also, my dream had always been to attend college. When I enlisted, the benefits were attractive and I loved learning. I knew that if I wanted to attend college the Army would be the best way to make that dream a reality. I continued working while waiting on my enlistment. Finally, the time came to follow through on my decision.

In January, 1987, I traveled to Ft. McClellan, Alabama for eight long weeks of Basic Training. I will never forget my feelings of panic knowing that I would not be able to see my family for two months. I hardly slept the night before my trip. I really felt the impact of my decision when I had to raise my hand and take the oath.

I will never forget that first day of Basic Training. We were all loaded onto a hot bus and driven to the site. As soon as the bus stopped and those doors opened, the Drill Sergeant ran in and started yelling for us to get off the bus quickly. As we exited, all of the drill sergeants were yelling at the same time for us to load our bags inside of the barracks.

"Move it! Move it!" they screamed in our faces. They were like angry dogs barking orders at us.

I had heard that the drill sergeants were very scary. I thought I had seen it all being raised by my father, but what I hadn't heard was just how fast they could make people move. I had never seen so many people load out of a bus with as much chaos in my life. One recruit even tripped on the step and literally road his bag down! For some reason, I guess we had not met the expectations of these drill sergeants because we were all ordered to start doing push-ups on the hot cement immediately after we put down our bags.

I couldn't understand it. What had we done to deserve immediate punishment? I soon learned that everything we did in Basic Training was preparing us to meet the mental, physical, and emotional demands of the Army. They had two months to break down and reconstruct decades of personal habits and beliefs to conform us to the Army's way of thinking and behaving. There was no time for orientations, icebreakers, and getting situated. We were prepared as though we were being prepped for the battlefield.

The drill sergeants put us into platoons and assigned a drill sergeant for the 8 weeks. They ran us ragged throughout the base camp, yelling and screaming for us to move to the next location as quickly as possible. Finally, the drill sergeants allowed us to take our gear in our rooms and explained to us what we could expect for the next two months.

"You will report for physical training at 0-five hundred hours" our drill sergeant barked in Soldier cadence. "Do you understand?"

"Yes Drill Sergeant!" we yelled so loudly. I believe we could be heard in the next state.

"Your PT Test will consist of a timed two-mile run, two non-stop minutes of push-up and two non-stop minutes of sit-ups. Is that understood?"

"Yes, Drill Sergeant!"

For the next 8 weeks, our days were filled with early morning workouts, weapons training, combat training, and field training, classes on rank structure, and more physical training. It was grueling. I knew what to expect and began working on my physical strength before going to Boot Camp. I was in the best shape of my life. I was running 4 miles a day, eating healthily and building my physical endurance. I had worked on doing push-ups and sit-ups, but I was still unprepared.

Most of the girls in my squad were in their 20's but, one recruit was in her 30's. She kept up with us for the 2 mile-run and we were always so proud of her for her courage to get through the training. This was a time to prove to the drill sergeants and ourselves just what we could accomplish. For the remaining seven weeks of basic training, we were disciplined on a regular basis and most of the time it was a group effort. I think the intent was to make sure that we all followed orders using peer pressure. The physical demands that are placed on the recruits can be extraordinary if you report to school unprepared.

I remember the first time I ran the 2 -mile event and finished. It reminded me of the many times, my grandmother warned me to get home before the streetlights came on; I would sprint home at breakneck speed to make it just in time before she locked her doors. When it came time to take the PT Test, we were all ready and excited to see what our scores would be.

The first time I did my PT test, I reflected on how strong my grandmother had been in my life. I knew that both she and my mother would be proud knowing that I completed Basic Training. Amazingly, the older recruit who we thought was keeping up with us was caught cheating. She had found a short cut through the route and wasn't aware that she was being shadowed. She had to re-run the entire route. She looked as though she was about to have a heart attack after running the full two miles. Eventually, she was sent home and lost out on an opportunity to change her life. This was the first time that I had witnessed just how seriously the Army standards could effect someone.

After completing and passing the horrendous PT Test, the Army had one more test. This one worse than the first. The final test was the 12-mile road march that had to be completed in "full battle rattle." That's what we called wearing full gear. Full battle rattle consists of: Full uniform, "FLC" (Fighting Load Carrier) which is a vest to hold canteens filled with water, ammunition magazines and flashlights, Bullet proof Kevlar vests and helmet, and a rucksack that weighed a minimum of 35 pounds. The weight of all of this gear can exceed 60 lbs.

If that wasn't enough, we had to hold our 7 pound, M16 Rifle the entire time. Most of the trip was uphill and it had to be completed within 3 hours. Imagine running 12 miles, non-stop, with a 10-year-old boy on your back while holding a seven-pound baby. This was clearly the most difficult part of basic training. To survive it, I reflected on the miles of walking with my mom to and from the store. I imagined in my mind that she was right by my side teaching me those lessons about patience. Man did I need patience on that march. There were times when I felt like the march would never end. I wanted to throw my M16 on the ground and scream, "I quit!"

But my mom was right there in my heart telling me to keep moving forward. Every step was another step closer. I could hear her teaching me about perseverance and overcoming the pain for the greater cause. My mother was hundreds of miles away, but she pushed me all the way to the finish line. When I finished the march, I collapsed from fatigue, but I was standing tall in spirit. I had overcome the greatest challenges of my life over the last two months and came out successful. I knew that I would be able to accomplish anything.

I will never forget the last day at Ft. McClellan; graduation day. There was nothing like the pride of the first day that I wore the Class A uniform and was recognized as an enlisted member of the service. It was one of the proudest moments of my life. Family members gathered and we assembled for our last formation together as a unit.

Chapter 5
The Active Duty Assignment

After completing Basic Training, I traveled to Ft. Sam Houston Texas, on March 17, 1987, for Advanced Individual Training (AIT). This is the training that enlisted Soldiers are required to take to learn his or her job. I chose to be a Dental Specialist (91E) when I sat down with my recruiter to decide my career path. I felt great about my decision to begin my military career in the dental field. I was receiving the type of training that would be useful after the life in the military.

The environment during this training was not as structured as Basic Training. We still had physical training in the morning, but nothing like what we endured during Basic Training. There was free time on the weekends to relax or study training materials. I was assigned to Ft. Hood, Texas, after graduating from this training and later was transferred to Ft. Stewart, Georgia. It was great being back in Georgia. I could see my family on some weekends. I needed my family's encouragement as I pursued my new career. The new challenges were not physical; they were emotional and psychological. It was the challenge of life as an African-American female in a male dominated world. I was a double minority: black and female. I did not let that stop me though.

I was determined to stand out. I went above and beyond the usual requirements of the mission. The headquarters held boards for Soldiers to earn recognition as: **Soldier of the Month**; **Soldier of the Quarter**; **Soldier of the Year**. Earning any of these titles would make a difference in my career, so my goal was to earn one or more of these boards. My focus became learning as much as I could to succeed at earning board status. I studied everything I could about my job and about being a Soldier. If there were seminars, I was present. I found articles; read every book I could get my hands on. I was determined to be twice as good as any Soldier.

The experience of preparing for the boards taught me that I needed to constantly challenge myself and take advantage of every opportunity to advance. The challenge of appearing before the board was like life: I was evaluated on the skills I brought to the table. I would constantly remind myself; I was not competing with anyone but myself. I would tell myself, "I am the daughter of a Vietnam Veteran. If anyone can succeed, I can."

Military is no different from civilian life. Barriers of gender equality and race are major factors. We all know the term "glass ceiling" — it's the invisible barrier hindering the advancement of women in their professional lives. However, there is a new term out there that may complicate the way we look at feminism: the "concrete ceiling."

Like glass ceiling, the term, "concrete ceiling" is a barrier for success. The difference between the two terms is that the concrete ceiling is a term specifically made for women of color. Why the need for a different term? A concrete wall reflects the barriers that women of color face more accurately.

Let's start by looking at the difference between the materials. While glass is tough, you can shatter it. You can see through it to the level above (and you know that there is something to aspire to). If you can see it, you can achieve it. Concrete, on the other hand, is practically impossible to break through by yourself. It's impossible to see through. This is what women of color face in the workforce; an often-impenetrable barrier with no vision of how to get to the next level.

(Racial Equity For Women of Color, the Glass Ceiling is Actually Made of Concrete April 19, 2016 Jasmine Babers).

The challenges of trying to break the concrete ceiling are tough and painful. I was finally able to understand what my father was going through when he would come home frustrated and fussing about the difficulties of advancing in his career. But I also remembered how he would wake up the next day and put on his uniform with pride and set out for the day to do it all over again. Each morning when I looked in the mirror, buttoning up my uniform, I saw my father smiling back at me, telling me to keep pushing.

Finally, I had my first experience of equality with my first assignment in the dental clinic. This dental clinic was located on post in an air-conditioned building with plenty of space for patients and staff. Working alongside military dentists was exciting. Every day was different and filled with new challenges. The dentists with whom I worked, trained the dental assistants in various aspects of dentistry: i.e. General Dentistry, Endodontics, Oral Surgery, Prosthodontics, and other aspects of patient care.

I enjoyed being part of the dental command. I felt like we were family. The style in which we worked really set our unit apart from the rest of the military. After a year working as a Dental Specialist, my supervisor recommended me for a training slot for the Preventive Dentistry Specialty Course. This meant I would return to Ft. Sam Houston. After graduation from this course in October 1988, I was trained and could treat patients as a dental hygienist for the Army. Additionally, I would place fillings alongside the dentists. This opportunity allowed me to work independently, learn patient care, and manage my own schedule.

On May 17, 1990, I graduated from the Primary Leadership Development Course (PLDC). This month-long course was designed to develop Soldiers and to prepare them for Leadership. During this time, it was required to obtain promotion to Sergeant. The Army (at the time) had a promotion system based on points. A Soldier was awarded points based on military education, civilian education, and physical training. The promotion board panel recommend points based upon what each Soldier earned and compared it to the monthly points needed to be promoted. The standard points for a Dental Specialist were higher than most jobs in the military. Although, I had military training, worked hard at my job, even earned

"Soldier of the Quarter for the first quarter 1990," I did not have a degree and knew that I would probably not be promoted to Sergeant without it. I decided not to reenlist for the Active Army or prepare a packet for the promotion board. My goal was to attend college full time and work toward earning my degree.

Chapter 6
The Reserve Assignment

On September 6, 1991, my full-time assignment ended. I immediately re-enlisted for the U.S. Army Reserves and moved to Atlanta, Georgia. As reservist I was able to serve the military on a part-time basis which meant that I had more time to enjoy my freedom and time as a civilian. I only worked one weekend out of the month and two weeks out of the year. The culture in the reserves was very different from full time active service; it was very relaxed.

I enjoyed living in Atlanta because of the many opportunities for African Americans. It was a city like no other. It was often called "The Black Mecca". I loved the culture, the restaurants, the clothes, and the people. One of those people in particular was my hair stylist, Calvin. He played an important role in my journey, not just as a stylist, but as a friend.

For African American women, the care of the hair is important especially when wearing the military uniform. As a black woman in a military leadership position, appearance and presence means everything. The many years of heat and chemical damage to my hair caused my need for a hair stylist that understood my hair and me. For a lot of black women, getting our hair done 'right' is a very important part of our lives. It builds our self-esteem and adds value to our lives. When we put on the uniform, having hair that is 'laid' puts the icing on the cake.

Calvin and his wife married very young and had worked together to build a very good life surrounded by a loving family. Calvin was very talented at being a stylist and had a large clientele. He was also a wise man of God, a visionary and a Veteran. Calvin started doing my hair and it grew into a lasting friendship. Calvin's shop is where I went to relax. But in a way, you could also say that Calvin's shop is where I went for reinforcements.

***

I was 27 and so excited about being in my new unit in the reserves. As a reserve hygienist, my training included working at a dental clinic treating military retirees. This assignment was great because it connected me with people who were older, seasoned, and had served their country with honor. These retired Soldiers were always so appreciative of the service that they received at the dental clinic. They would tell the most extraordinary military stories from the beginning of their careers. It was precious to me because many of them talked a lot and gave long speeches just like my dad. Listening to them took me back to my childhood.

I truly appreciated it when they shared this part of their lives with me. I never took it for granted. If you understand the life of a Soldier, you'll understand their experiences are hard, traumatic, and held very close. Many of these valiant Soldiers lost best friends, loved ones and companions. So to be a sounding board for these men and women was a distinct honor. Talking was a way for these retirees to heal, so I listened. Later in my career, I would be able to use my experience of being a sounding board for these veterans when I became a veteran counselor.

***

On August 27, 1993, I was promoted to Sergeant E-5. This was the leadership opportunity I had been waiting to receive. My father retired a Sergeant E-7 after 20 years of service. His example, sacrifices, and commitment paved the way for his daughter to achieve at a faster rate. I was not going to let his efforts go in vain. I was determined to go even further.

I had to really take my career seriously if I wanted to be advanced further to a senior level Non-Commissioned Officer. I knew I was going to have to push myself harder than I ever had. In order to reach higher ranks, I would have to receive an excellent evaluation report. A non-commissioned officer is a military officer who has not earned a commission from the president. Non-commissioned officers usually obtain their position of authority by promotion through the enlisted ranks.

The Non-Commissioned Evaluation Report (NCOER) is used to evaluate a person in a position and recommend them for promotion to the next rank. This document is placed inside of a promotion packet and sent before a panel of board members to evaluate Soldiers. This document reports every detail about a Soldier, including: if he or she is meeting height and weight requirements and if the he or she is successfully completing the Army Physical Fitness Test. Promotion board members could also use this report if a Soldier was not ready to be promoted.

I was burning the candle at both ends. I was working out daily to keep physically fit for the PT Test and also working on correspondence courses. Also, I continued to treat patients in the clinic and other duties as assigned. I worked at stocking merchandise for my civilian job, as well as enrolled at a Technical College in Atlanta, Georgia. A third shift position posted to unload merchandise. I applied, worked at night, and attended school during the day.

I graduated as a Medical Assistant on September 27, 1993. Soon after, I got a job working at a private dental office. I really enjoyed these times that allowed me to work closely with the patients, learning about dental procedures, cost and billing. In December 1995, I got a job working as a Patient Treatment Coordinator, for a dental group. I enjoyed this position because it involved working with insurance and patient care. I had an advantage working in this position because of my experience working in the back office as an assistant. It afforded me the opportunity to educate the patients on their dental procedures as well as explain their insurance costs.

ser·geant
/ˈsärjənt/
noun
noun: sergeant; plural noun: sergeants

A noncommissioned officer in the armed forces, in particular (in the US Army or Marine Corps) an NCO ranking above corporal and below staff sergeant, or (in the US Air Force) an NCO ranking above airman and below staff sergeant.

Chapter 7
Disability

Life was going well for me. All my plans were falling in line just as scheduled. I was advancing in my career and getting nods of approval from my superiors, but little did I know, I would be facing one of the greatest challenges of my life. Greater than any physical challenge I had ever experienced.

In March of 1996, our unit attended a weekend training to qualify with our weapons. The day was going well. It was a nice warm day in the summer, but not too hot. This was a perfect day. I was firing off rounds from the semi-automatic weapons and joking around with my unit. As the day went on, I started feeling a nagging pain in the back of my head. It was warm, so I figured I was getting dehydrated. Also, the loud gunshots weren't helping. Midway through the training, my head was in excruciating pain. I tried to walk away but before I knew it, I was on the ground. I thought I was going to die.

Members of my squad rushed to my aid and took me to the emergency room. The doctors ran a full examination on me.

"Did you hit your head at any point?" They asked.

"No." It was so painful to even say a one syllable word.

"How long has your head been hurting?"

I just winced and got out as many words as I could.

"Not…. Long."

"We need to perform a spinal tap" the doctor told the nurse.

"Tammie, we need to perform a spinal tap. We need you to sign the release to treat you."

When a patient comes in with head pain, doctors will sometimes perform spinal taps to diagnose any signs of infection, bleeding in the brain, and to measure the pressure around the brain.

I scribbled my name on the paper, desperate for them to do anything to take the pain away. If they had asked me to give them permission to remove my head from my body, I would have signed that paper just as quickly. I waited for the neurologist for what seemed like hours and when she came, the procedure seemed to take forever. I had experienced a lumbar puncture when I had my daughter, and it didn't take this long. The neurologist was back there pushing and prodding. I just wanted to lie down. Finally, she was able to extract a sample of my spinal fluid for testing.

I eventually fell asleep and the pain was going away. The next day the doctor felt I had progressed enough to go home. The pain was gone but I was exhausted. My Aunt Dorothy took me to her house to rest. A few days later, I was having difficulty standing. It felt like someone had my back in a vice grip. Each day the pain got progressively worse. Aunt Dorothy couldn't take it anymore. She helped me get in her car and was taking me back to the hospital. I was terrified to go back. I didn't know what they did to me, but clearly the back pain I was experiencing was done by the doctors. When we got to the hospital, I was screaming for Aunt Dorothy to turn around.

"Please, take me back home Aunt Dorothy! Don't let these people get me."

The nurses came out and carried me inside. I was in too much pain to resist. Luckily, there was no need for another spinal tap. A shot of strong pain medicine was all I needed. The shot helped but when it wore off, I was miserable again. At one point, I lost the ability to walk. This went on for months. Reserve Soldiers are not entitled to the same medical care as their active duty counterparts. Fortunately, the Army was able to take care of my needs through short term disability.

In addition to the physical pain, I was experiencing mental and emotional pain too. I was growing depressed about the fact that I may never reach my goals. My father had fought so hard so that I could make the strides I was making, and now I was disabled. In my mind, I was losing everything, and I was letting my daughter down.

I remember thinking "all of this hard work was for nothing." In my mind, the only thing that mattered was complete success. There were also financial ramifications to not meeting a full 20-year active duty career with the Army. There is a threshold a Soldier must reach in order to receive full retirement benefits. I had fought for every inch of the 10 years of progress I made in the army, but 10 years would not afford me anything more than bragging rights. My career had hit a brick wall.

Aunt Dorothy was my rock during this time. She was a very wise woman who held and lived very strong Christian values. She served on the Mother's Board at her church and taught in the education system for over 30 years. She and her husband raised four children in their home in Atlanta, Georgia. Aunt Dorothy would give me words of wisdom and encouragement. She would speak words of life over me. She was so amazing. She fed me, she drove me to the hospital, prayed for me, believed by Faith for my healing.

My Aunt Dorothy is my Super Hero. She is to this day because of her faith and ability to trust and rely on God. She nursed me back to health. After months of disability, I was finally able to walk, run and return to my goals.

CHAPTER 8
JAPAN

In April 1997, I was 100% healthy and ready to return to the Reserves. I called my Captain and asked about summer training.

"Well, looks like you just missed our Hawaii training."

"Aww man! Do you have any other opportunities?"

"We have a training in Japan next month."

"I'll go."

"We can cut your orders for next month."

I would be on a two-week training assignment as a dental assistant in Japan. I told my daughter about the opportunity, and to my surprise, she was excited for me. That was all I needed. I was thirty-three, single, and excited about this assignment. I had never traveled this far from home alone.

The military has a policy and regulation called *The Family Care Plan.* This packet of documents is put in place for a military member separated from their family. It is a detailed plan to inform family members, your unit, and others what your wishes are in terms of the care of your children, insurance, funds, and other issues concerning the Soldier. A Family Care Plan should be developed whether you expect to be deployed or not. In fact, units will require you to develop a formal Family Care Plan.

Taking care of the considerations early will help you and your family be prepared for any type of separation.

When you report to a new duty station, you have the responsibility of finding good health care, schools, and medical providers that you trust. It is not easy for women or men who serve to leave children behind when called on a mission.

I processed my Family Care plan and set off to Japan. The flight was about 13 hours, but I didn't mind. It was a new adventure. The plane landed in Tokyo. From there, I took a bus to my duty station. Many people don't know the opportunities the Army affords Soldiers for career development. I stayed in a hotel on the military base (all expenses paid) which was located directly across from the dental clinic. I worked with dentists on base along with other dental assistants and staff. I learned more about dentistry in a different culture.

There was a dining facility on base with plenty of activities for Soldiers. On the weekends, I traveled to the city for shopping, eating at restaurants, and touring the city. I enjoyed meeting people from Japan. They were kind and had very traditional ways. While visiting in the city, I got a chance to see a wedding with all the customs and courtesy. This time in service was amazing for me, it made me miss serving in the military on a full-time basis.

When I returned to the states, I put in an application for the Active Guard Reserve (AGR) Program. This program accepts members of the Reserve Component and allows them to serve alongside the Active Duty component full time. When I made the decision to apply for the program, I decided to retrain in the Personnel field, due to the many opportunities for advancement. This meant that I would have to attend an active duty school and meet the many physical and academic requirements required to be successful.

After being accepted into the program, I was informed that I would receive an order placing me back on Active Duty. When I received my full-time assignment orders, I was surprised when I looked at the actual date of the order. It was May 14, 1998, the date was my birthday. I received this as a sign that God was going to make a way for me. He had broken down the brick wall, and I was free to keep moving towards my goals. I was assigned to Ft. Jackson, South Carolina, for eight weeks for training, and then on to my first assignment in Nashville, Tennessee. I resigned from my civilian position, gave my notice to my apartment manager and moved my furniture to storage. I hated to say goodbye to my family, friends, church, and Soldiers in my reserve unit. There was no turning back, after I received my orders.

I reported to Ft. Jackson, South Carolina, in June 1998, one of the hottest months of the year. I had to be prepared mentally and physically for the days ahead. The PT Test with height and weight was given within the first few days of reporting to the school. I had most of my clothes in my car ready to set up in my new home in the barracks at the school. I had my mind set to get through the physical and academic requirements. This move meant, if I failed any portion of those requirements, the Army would have to terminate my orders. I would have been without a place to live. Once again, I was laying it all on the line. It was time to give my best, push my hardest, with a NO QUIT attitude!

CHAPTER 9
NASHVILLE

In August 1998, after graduation from Ft. Jackson, South Carolina, I reported to Nashville, Tennessee as a Personnel Services Sergeant. I was 35 years old and looking forward to my new job in the military. I was excited to make this adjustment to working full time in the service. I knew that if I needed to visit friends and family, I could easily make the four-hour drive back to Georgia.

Within days of moving to Nashville, I visited Lake Providence Missionary Baptist Church, and met the Pastor Reverend H. Bruce Maxwell, a native of Nashville who had a passion for preaching the Word of God. He was elected to serve Lake Providence on September 10, 1976. He and his wife both love God and have faithfully served the community of Nashville for many years. I began to attend services on a regular basis along with weekly Bible Study. Eventually, I joined the church and attended the new member's class and that is where I met Deacon Lee Allen. He was the head of this ministry within the church and he treated all the members like family. I felt very hopeful about my new church, position in the Army, and my community.

On August 22, 1999, after just a year of living in Nashville, I received some devastating news that my father had passed away suddenly. He was 71 years old. He had died at the Veteran's hospital in Texas. His body was flown to

Georgia for the funeral. This is the first time I had lost a family member so close to me. I tried to accept his passing but it was difficult because I never got a final speech from him. I had no closure. In the days before his funeral, I tried to come to grips and accept that he was gone. The reality of it all really did not seem to register in my mind. The challenge seeing my father for the last time was tough.

After the funeral, my depression and self-doubt set in deep. Could I continue in the service? Did I even want too? Did I have what it took to be successful? I had never experienced grief like this and knew that I could not do it alone. I continued praying and asking God for strength. My daughter gave me something to fight for when I wanted to give up. Early in her life, my daughter and I spoke of her becoming a pediatrician because of her love for children. She has always been mature for her age and very goal oriented. I prayed that whatever decision she had made that God would bless her. She was a junior in high school, living in Texas with her father. Although I was in pain, I was excited about the opportunity to help our daughter plan for college and provide support for her future.

I received some pictures of my dad during the time of his service. Also, I received his certificate of appreciation from President Nixon. When a Soldier retires, he or she receives a certificate from that current president. Getting those items helped me with my healing process. When I attended services at my church, I spoke with other members who had suffered loss. My ability to lean on my true strength, in God, was vital in being able to continue with my journey. I thank God for my Pastor and my church family. Because of those individuals, I knew that I would not suffer this loss alone.

Soon after my father passed, my family and I assisted my mom in investigating her benefits to ensure that she would be taking care of. The military has a way of taking care of the family members who are left behind after the service member passes away. These survivor benefits are helpful and can ease the financial burden that often falls on the family. When I was in the military, I was always moving at a constant pace, learning my job, training, and keeping up with the daily demands of the mission. There was usually not a lot of time to process my feelings. I have grown to believe that feelings have a way of creating pockets within the heart. During this time, I had multiple stresses in my life that I had to deal with, including: being a Soldier, being away from my family, getting used to a new culture, and the death of my father.

Cumulative stress is a common experience for people who work in chronically stressful situations. It results from an accumulation of various stress factors such as heavy workload, poor communications, multiple frustrations, coping with situations in which you feel powerless, and the inability to rest or relax.

I had to find of way of releasing my feelings about losing my dad, so I made an appointment for counseling to speak with a trusted source. This is when I was introduced to "Talk Therapy". It was during one of my counseling sessions, I became fascinated with the field of psychology. My therapist recommended a few books, those taught me about the five stages of grief, "denial, anger, bargaining, depression, and acceptance". (Elisabeth Kubler-Ross, David Kessler, *Finding the Meaning of Grief through the Five Stages of Loss*). He said that my feelings were normal. I was experiencing the stages of grief. As I started to experience the stages; I knew that I wanted to know more about the field of psychology. I was introduced to topics: Psychotherapy, the significance of Therapy Sessions, Psychotherapy and Medication, Cognitive Behavioral Therapy, or Dialectical Behavior Therapy. These changed my outlook on my future.

---

*www.oncologynurseadvisor.com/home/the-total-nurse/cumulative-stress/

CHAPTER 10
UNDERSTANDING PSYCHOTHERAPY

In my opinion, this chapter is one of the most important chapters of the book. Yes, I want everyone to know that they can accomplish any goal to which they set their minds. I want civilians to know about the benefits and opportunities the U.S. Army provides. But I really want my fellow Soldiers to understand the importance, impact, and absolute necessity of therapy.

What is Psychotherapy?

Psychotherapy, or talk therapy, is a way to help people with a broad variety of mental illnesses and emotional difficulties. Psychotherapy can help eliminate or control troubling symptoms so a person can function better and can increase well-being and healing.

Problems helped by psychotherapy include difficulties in coping with daily life; the impact of trauma, medical illness or loss, like the death of a loved one; and specific mental disorders, like depression or anxiety. There are several different types of psychotherapy and some types may work better with certain problems or issues. Psychotherapy may be used in combination with medication or other therapies.

Therapy Sessions

Therapy may be conducted in an individual, family, couple, or group setting, and can help both children and adults. Sessions are typically held once a week. Both patient and therapist need to be actively involved in psychotherapy. The trust and relationship between a person and his/her therapist are essential to working together effectively and benefiting from psychotherapy.

Psychotherapy can be short-term (a few sessions), dealing with immediate issues, or long-term (months or years), dealing with longstanding and complex issues. The goals of treatment and arrangements for how often and how long to meet are planned jointly by the patient and therapist.

Confidentiality is a basic requirement of psychotherapy. Also, although patients share personal feelings and thoughts, intimate physical contact with a therapist is never appropriate, acceptable, or useful.

Psychotherapy and Medication

Psychotherapy is often used in combination with medication to treat mental health conditions. In some circumstances medication may be clearly useful and in others psychotherapy may be the best option. For many people combined medication and psychotherapy treatment is better than either alone. Healthy lifestyle improvements such as good nutrition, regular exercise, and adequate sleep can be important in supporting recovery and overall wellness.

Does Psychotherapy Work?

Research shows that most people who receive psychotherapy experience symptom relief and are better able to function in their lives. About 75% of people who enter psychotherapy show some benefit from it.

1. Psychotherapy has been shown to improve emotions and behaviors and to be linked with positive changes in the brain and body. The benefits also include fewer sick days, less disability, fewer medical problems, and increased work satisfaction.

With the use of brain imaging techniques researchers have been able to see changes in the brain after a person has undergone psychotherapy. Numerous studies have identified brain changes in people with mental illness (including depression, panic disorder, PTSD and other conditions) as a result of undergoing psychotherapy. In most cases, the brain changes resulting from psychotherapy were like changes resulting from medication.

2. To help get the most out of psychotherapy, approach the therapy as a collaborative effort, be open and honest, and follow your agreed-upon plan for treatment. Follow through with any assignments between sessions such as writing in a journal or practicing what you've talked about.

Types of Psychotherapy

Psychiatrists and other mental health professionals use several types of therapy. The choice of therapy type depends on the patient's illness and circumstances and his/her preference. Therapists may combine elements from different approaches to best meet the needs of the person receiving treatment.

Cognitive behavioral therapy (CBT) helps people identify and change thinking and behavior patterns that are harmful or ineffective, replacing them with more accurate thoughts and functional behaviors. It can help a person focus on current problems and how to solve them. It often involves practicing new skills in the "real world."

CBT can be helpful in treating a variety of disorders, including depression, anxiety, trauma related disorders, and eating disorders. For example, CBT can help a person with depression recognize and change negative thought patterns or behaviors that are contributing to the depression.

Interpersonal therapy (IPT) is a short-term form of treatment. It helps patients understand underlying interpersonal issues that are troublesome like unresolved grief, changes in social or work roles, conflicts with significant others, and problems relating to others. It can help people learn healthy ways to express emotions and ways to improve communication and how they relate to others. It is most often used to treat depression.

Dialectical behavior therapy is a specific type of CBT that helps regulate emotions. It is often used to treat people with chronic suicidal thoughts and people with borderline personality disorder, eating disorders and PTSD. It teaches new skills to help people take personal responsibility to change unhealthy or disruptive behavior. It involves both individual and group therapy.

Psychodynamic therapy is based on the idea that behavior and mental well-being are influenced by childhood experiences and inappropriate repetitive thoughts or feelings that are unconscious (outside of the person's awareness). A person works with the therapist to improve self-awareness and to change old patterns so he/she can more fully take charge of his/her life.

Psychoanalysis is a more intensive form of psychodynamic therapy. Sessions are typically conducted three or more times a week.

Supportive therapy uses guidance and encouragement to help patients develop their own resources. It helps build self-esteem, reduce anxiety, strengthen coping mechanisms, and improve social and community functioning. Supportive psychotherapy helps patients deal with issues related to their mental health conditions which in turn affect the rest of their lives.

Additional therapies sometimes used in combination with psychotherapy include:

Animal-assisted therapy – working with dogs, horses or other animals to bring comfort, help with communication and help cope with trauma.

Creative arts therapy – use of art, dance, drama, music and poetry therapies.

Play therapy – to help children identify and talk about their emotions and feelings.

More Information

Academy of Cognitive Therapy

American Association for Marriage and Family Therapy

American Psychoanalytic Association

American Academy of Psychoanalysis and Dynamic Psychiatry

Association for Behavioral and Cognitive Therapies

References

American Psychological Association. Understanding psychotherapy and how it works. 2016. http://www.apa.org/helpcenter/understanding-psychotherapy.aspx

Karlsson, H. How Psychotherapy changes the Brain. Psychiatric Times. 2011.

Wiswede D, et al. 2014. Tracking Functional Brain Changes in Patients with Depression under Psychodynamic Psychotherapy Using Individualized Stimuli. PLoS ONE. 2014. http://journals.plos.org/plosone/article?id=10.1371/journal.pone.0109037

Physician Review By:

Ranna Parekh, M.D., M.P.H.

Lior Givon, M.D., PH.D.

Chapter 11
The Personal Trainer

I remember the first day I saw him. He was so visually remarkable. I don't think I would have missed seeing him, even if I was blind in both eyes. He was tall, brown skinned, had almond shaped eyes, and a very welcoming smile that exposed a gap in his front teeth. He had a very exotic look about him. I could tell he was not from the states.

The second time I saw him, I was on my way to class. My phone battery had died, and I was very concerned about reaching my mom. When he saw me, I guess he noticed how distressed I looked.

"Can I help you?" When he spoke to me, I was mesmerized by his strong accent. I appreciated his sincerity and willingness to help me.

"Yes. Can I use your phone? My battery is dead, and I really need to contact my mother."

He was very kind. I could tell by the way he stared that he was attracted to me.

After reaching my mother, I was so relieved and thankful, I gave him a big hug. Well, maybe I wanted to wrap my arms around him regardless of the phone, but this was the perfect opportunity. He smelled like delicious pound cake.

"Thank you so much. You're a life saver. What's your name?"

"My name is Innocent." He said with that blazing smile.

"Innocent. Really?" I asked. I figured he was joking with me.

He laughed at my curiosity.

"I am very serious. Where I'm from, Innocent is a common name."

"Oh yeah, and where are you from?" It felt so easy to talk to him. I just felt comfortable with him.

"I'm from the Igbo tribe of Nigeria."

The Igbo people are one of the largest ethnic groups in Africa. The language of the Igbo people is Hebrew and there is speculation that they may be the original Israelites. Researchers and scholars found that Igbo Jews practice Judaism so authentically, that they either adopted Judaism when it was first started or they are descendants of the Jews of Bilad el-Sudan.

Innocent was a jack and master of many trades. He was a professional boxer, musical entertainer and part time security guard. He shared his love for God and that he was very close to his family and really enjoyed keeping in contact with them in Nigeria. I appreciated that we both had a strong a connection to family.

After many conversations, he finally asked me out on a date. I wanted to ensure that we both were comfortable on our first date. I didn't want to go to some snobby, fancy restaurant. I just wanted to get to know him as he truly was. When he came to pick me up, he was in his uniform. I looked at him and said, "If we go out, you can't wear those big shoulder pads." When I reached my hand out to touch him on his shoulder, I realized that he was not wearing shoulder pads. There he was standing before me with shoulders as wide as the sea. He was as strong as those shoulders looked and just that fine.

Over the next few weeks, we started spending time together. I found out he enjoyed cooking. He was great at it. The first dish he made had rice, chicken, beef with an amazing tomato paste sauce. He purchased his seasonings from the neighborhood African store. He made a dish called "fufu" with his meals (ate it with his hands). The dish was served with every meal and very common tradition in Nigeria. He told me, "a day without fufu was a wasted day."

Later, I found out he really enjoyed working out. He was great at that too. We would run for miles, do push-ups and sit-ups. I was surprised he could keep up with me. I think he was allowing me to keep up with him. It came time for me to move to another apartment within the city and I needed some help. I asked him, if he would help me move some of the boxes while I was at work. I expected that he would help by moving a few boxes out, to lessen the work load when I got home. When I got off work, I arrived home very surprised to find he had single handedly moved everything out of my apartment into my new place. He displayed actions of kindness from the very beginning which was another attractive quality.

We would attend Nigerian events together. I really enjoyed meeting his friends and family. I enjoyed those amazing African dishes that were served at these events. Soon, I learned how to dance and dress in the traditional attire and even speak some of the language. As we grew closer, I grasped he was an extremely patient person. He held a lot of respect for the military. I really enjoyed the fact that I had someone in my life who supported me. This was a man with whom I could spend the rest of my life.

Chapter 12
The Second Active Guard Reserve Assignment

On August 16, 2002, I got remarried in Minnesota, during my second AGR assignment. I was 38 years old, excited, and nervous at the thought of living so far away from my Georgia roots. I knew the adjustment would be different for me. However, I was up for the challenge. Minnesota was not as culturally diverse as a lot of places I lived. I initially had issues finding a church, soul food restaurants, hair care products and stores that I liked.

Soon after moving there, I found an African Methodist Episcopal (A.M.E.) church. I always had my mom's voice in my head reminding me to "find a church home and trust in God." I met a very dynamic pastor and wife team. The ministry held corporate prayer with the members during the service. The prayer reminded me of the way we prayed in my church growing up. This church provided worship and gave me stability as a Soldier in a new place.

The culture in Minnesota was very different from the south, where I had grown up. I, being a military brat, soon made friends. I learned how to live away from my family. The good thing about living in Minnesota was that my husband had many friends from Nigeria, and they lived close by. We attended many African events held there, such as weddings and parties. I learned how to prepare some of the Nigerian dishes.

Minnesota was one of the coldest places I had ever lived. There were days that I would get up, go outside and want to move back to Georgia! It was also a challenge for my husband because he had always lived in a warmer climate. It was so cold, people would have to use warmers for their cars just to get them to start in the mornings. At times, during the coldest months, you could literally throw water in the air and watch it freeze.

The mental adjustment was challenging for me. There were times it remained dark in Minnesota much of the day. It would affect my mood. I was diagnosed with Seasonal Affective Disorder (SAD). The doctor advised me to cut on all the lights in the house, in order to simulate sunlight. I decided to keep busy by focusing on completing my Associates Degree, as well as making progress in my career. I received a free laptop issued by the EArmyU Program, before leaving Nashville. This was a program offered to service members to complete their education online. The one requirement to remain in the program was to complete 12 hours of college courses I was determined to take advantage of every training opportunity.

In June 2003, I attended training in Germany for two weeks to provide logistics support to another unit. Then, August 2003, I attended the Total Army Instructor Training Course in Wisconsin. This was a course that provided Soldiers an experience of giving classes as an instructor.

The balancing act of work, education, a new assignment had its challenges, yet was rewarding.

During this time, my daughter was attending college in Texas. She was in her sophomore year making good grades and even competing in pageants. I was thankful she was doing well on her own. The college experience gave us a lot to talk about. I am grateful for the bond we shared.

After many months of working in Minnesota, I was promoted to Staff Sergeant E-6, on the first of November 2003. I was so excited about having more responsibility and a new role in the military. I was transferred to another unit in Minnesota. This allowed me to learn about transferring Soldiers to other units, creating documents for promotion, and assisting Soldiers that were leaving the military.

In April 2004, I returned to Ft. Jackson, South Carolina for the Basic Non-Commissioned Officer Course. This course was probably the most challenging military course because of the rigorous schedule. There were Soldiers there from all over the country. We were taught how to conduct military classes, field operations, and how to properly conduct drill and ceremony. I graduated from this course in June 2004. There were many classes that were taught that prepared Soldiers for leadership. During this time, I remained enrolled in online classes, along with military correspondence courses. I completed my bachelor's degree in June 2005.

A Staff Sergeant is a non-commissioned officer (NCO) in the United States Army usually placed in command of a squad of 9-10 Soldiers. In rare situations, a Staff Sergeant may be placed in command of a larger unit such as a platoon comprising of two to four squads containing anywhere from 16 to 50 Soldiers.

Chapter 13
The Mom Soldier

My journey took an unexpected turn in 2005. My husband and I were getting accustomed to life in Minnesota. We had been married for three years, and we were loving life. My daughter was grown and independent. She had just completed her Master of Social Work degree and began working in her field. I was really feeling like I was on the downhill, coasting side of life. I was in a great relationship, my career was progressing just as I wanted, and my daughter was on her way to great things. How much better could it be?

My husband had an idea to make it better:

"Babe, I want a child," he said one day.

Screech!! I hadn't considered the idea before. I mean, to be honest, there were a couple of major obstacles that had to be considered. First of all, I was over 40. Secondly, my tubes were tied. However, I loved this man and having a child with him would have made me just as happy, so I started to research the possibilities of getting pregnant. I was referred to a website called *Pregnant Again*. I decided to make the phone call.

I learned all about the procedure to untie my tubes so that I could conceive again. The call was reassuring, and I was comfortable about making the initial appointment. In May of 2005, three days after my 41st birthday, I met with

Dr. William A. Greene. He performed the surgery, but he did not have good news.

"Mrs. Otukwu, unfortunately, we could only repair the left tube."

My hope was crushed. It was difficult to believe that this would work without me being in 100% tip top shape to conceive. After all, I was older. I wasn't some 20-year-old. The chances of getting pregnant decrease as you age. However, my grandmother had taught me about the power of prayer, so I went to church every week for corporate prayer. I was praying for a miracle.

"Lord. I know that You can do all things. I need a miracle if we're going to conceive. Lord, I don't want my baby to be unhealthy or have any problems due to my age. Please, Lord, bless me to have a strong and healthy baby."

I felt better after that prayer. I decided that I would trust God and stop trying to make things happen. As my grandmother would tell me when it would storm: "have faith and know that God is doing His work." I decided I would stop worrying about getting pregnant and focus on training for my upcoming PT Test and classes in graduate school. I still had goals to achieve. I knew that God would handle the things I could not, but I still had to take care of the things I could control.

Three months after the procedure and one day before my PT Test, I woke up not feeling my best. I wondered if I was just worried about the PT test. *Wait! I haven't had my cycle yet.* I thought. I had been so busy with classes and training, I hadn't even thought about getting pregnant.

I took a pregnancy test and to my surprise, the test was positive. I was overwhelmed with joy. I couldn't believe it.

***

My pregnancy was going well. The ultrasounds and tests for women giving birth over the age of 40 had all come back good. I didn't have issues with high blood pressure, and there were no signs of the baby having any problems. I praised God that He had answered my prayers. It was a miracle that I was able to get pregnant, but even more to have a healthy, problem-free pregnancy. My appetite had increased, and I craved African food. I ate constantly, even during my work hours. I was careful to not overeat. I knew after having the baby, I had to meet the Army weight standards and continue to successfully pass the PT Test. I went to all my appointments and received news from my doctor that things were going as expected.

I was also preparing to get my records ready for the next promotion board. This was the first promotion board I presented my records to while I had been pregnant. This meant that I would not have information about my PT Test, height or weight on my (NCOER) to be evaluated against other Soldiers. I was exempted from these standards because of the pregnancy.

"Lord, You know I don't have everything I need, but I pray that you touch the hearts of the board on my behalf."

I remained busy and worked all the way up until the day that I went to into labor. On June 12, 2006, we were so overjoyed to meet our healthy baby boy.

My husband was an amazing proud dad. He immediately took on his role of changing diapers, feeding, and making sure that we were both getting rest. He worked a third shift position and took care of our son during the day for the first year of his life. This meant that we did not have to worry about the stress of finding daycare. Having a supportive spouse was a very important aspect of military life especially during this time because I was still in graduate school. I had to complete my studies while working full time after my medical leave ended.

***

The time finally came for the promotion board results to be published and I was pleasantly surprised that I had made the list. I was going to be promoted to Sergeant First Class! This was the height of my father's career, and I was on the doorstep of making my goal. On November 30, 2006, I completed my Master of Management Degree.

Being a wife, mom, graduate student and a Soldier was truly a juggling act. Even with perfect time management, it was very stressful. Women in the military face unique challenges. We are often the primary caregivers for the children and are still expected to fulfill our military responsibilities. Yet, even though I didn't have my height, weight, and PT test for the board review, the board still promoted me. As my grandmother would say: "God was doing His work."

Sergeant first class

Sergeant First Class (SFC) is the seventh enlisted rank (E-7) in the U.S. Army, ranking above staff sergeant (E-6) and below master sergeant and first sergeant (E-8), and is the first non-commissioned officer rank designated as a senior non-commissioned officer (SNCO).

Chapter 14
The Conversation

It was Tuesday, July, 10 2007. I will always remember that day. I received the phone call from my mother's doctor.

"We are going to have to remove her leg this week."

I will never forget the feeling of panic I had. I couldn't even imagine the horror she was feeling at the idea of losing a limb. I was overwhelmed with my own horror and I wasn't the one who would be losing the leg. The thought of my mother having to face such a major surgery was too much. I could not believe this was happening to her. My mother was the nicest person I have ever met. She had faced so many struggles in life, including the death of two children: my sister's Pamela and Kathryn. My sister and her husband had been taking care of mom in their home. We were not sure what we were going to do in terms of her post-surgery care. My mother was also diabetic which made her condition even worse.

So often in the military, Soldiers become accustomed to handling everything that comes their way. They usually don't complain. They just accept each mission by carrying it out with whatever skills and abilities necessary. At the time, I could not push back my feelings or even try to pretend that things were fine. It was as if my world stopped. I could not think of moving forward. I could not think of any words to say to my mother to make everything better or fix this situation for her.

I had to speak to her, even if I stumbled on every word. While conversing with her, I remember trying to not speak out of my fear. I did not want her to know my out of control thinking was not in Faith mode. I tried to tap into her feelings by asking about how she would prepare for going into the surgery. Our conversation was short, she would not allow me to give into my fears. I will always remember her words during this conversation. I told her I was thinking about coming home to take care of her.

"No!" she shouted. "What you are going to do is stay in the Army and earn your retirement! Tammie, I have prayed all about that, can we talk about something else."

My mother had made her peace with the situation and with God. She had no fear. Even at the age of 42, my mother was still teaching me how to trust in God through her example. My mother was moved into a nursing home after the surgery. When I made my first visit to see her after the surgery, I was very pleased to see how well she was doing. In fact, her health improved very well after the surgery.

I became stronger after my conversation and visits with my mom. Perhaps, it was seeing my mother's strength that motivated me to continue my goals. I was committed to making sure I did as she had told me and concentrated on my retirement. This letter entitled, "Little Did I know" was birthed from my reflections after my mother's surgery. She inspired me beyond measure by living her life to her fullest potential.

Little Did I Know

When I was growing up: I used to often wonder why my mom never worked outside of the home.

Little did I know… that God was using her to be there for me and my siblings, to nurture us, so that we would have the strength and courage to face life's challenges.

When I was growing up: I used to wonder why my mom never learned to drive. We used to walk for miles and miles to the store; to purchase groceries for the family.

Little did I know…that God was using my mom as my personal trainer, getting me into shape to be able to withstand the rigorous endurance tests in the military I would one day have to sustain.

When I was growing up: I used to wonder why my mom would always prepare my dad's military uniforms by hand; not taking them to the cleaners like other wives.

Little did I know…that God was using my mom to show me what an awesome blessing it is to serve others.

Now that my mom is in the hospital and going through her struggles, I know that God is once again using her to show me He is faithful and will indeed make a way.

Loving You Always,

(Your Daughter) Tammie

CHAPTER 15
TIME TO SLOW DOWN?

My son was two years old when I had to travel to California to attend Advanced Non-Commissioned Officer Course (ANCOC). This was a course for senior leadership. This trip was harder than the others. I had never been away from my son for that length of time.

When I had to be separated from my daughter, it was for the greater purpose of making something of myself and being able to provide for her. But now, I had achieved my goals. I had enjoyed many accomplishments both personally and professionally. Leaving my family for weeks at a time was not as high on my priority list. My husband took care of our son during this period, and even though I really enjoyed the opportunity to see California for the first time, I realized something was changing in me.

The trip to California also gave me time to visit and spend quality time with my daughter and her fiancé. She had started planning her wedding. I had dreamed of this opportunity to sit down with my daughter and pick out wedding dresses, wedding locations and decorations. It was all that I imagined. All of the years of sacrifice, punishing days of physical training, and even fighting through physical disability seemed worth it just to be here with my daughter and be capable and able to help her see this special time in her life.

The time had come for the nuptials. I traveled to Maryland to attend my daughter's wedding. It was such an exciting time. The families came together and blended beautifully. I knew that my daughter and her husband would create a wonderful life together. I was somewhat nervous about being a mother in-law. Would I be able to guide them in marriage and be a strong model like the women in my life? The one thing I knew for sure was that my son in-law was a strong man of God. I believed he would always take care of my daughter.

During the first week of November 2008, about a month after returning from the wedding, I went to the emergency room after experiencing pain in my side. The doctor treated me with pain medication and said that I had a cyst. When I returned home, I sat on the couch and was getting ready to take a nap and experienced an awful pain in my left leg. My leg was throbbing and felt hot to the touch. Since I had just returned from the doctor, I saw no reason to go back. I called a friend who had experience in the medical field and told her what was happening. She said, "You need to go to the doctor, it might be a blood clot". I called my doctor who instructed me to come into the office immediately.

"Tammie" she said with panic on her face. "We have to get you to the hospital right now!" I was rushed into the operation room as quickly as possible to remove the clot. I was in the hospital during election night. A young Barack Obama was making history to be the first Democratic candidate for President. Could this really happen? After 231 years of existence, would I really be blessed to see the age when America would elect a Black man as president? Would I have the honor to serve for a Black president and call him my Commander in Chief? I was so excited to be a Soldier in the military during such a historic event.

Tears of joy slid down my cheek as I watched President-elect Barak Obama step to the podium in front of 240,000 cheering Americans in Chicago, Il.

"If there is anyone out there" he began with his cool, deliberate, drawl "who still doubts that America is a place where all things are possible; who still wonders if the dreams of our founders is alive in our time; who still questions the power of our democracy; tonight is your answer."

I was talking on the phone with Kim, my cousin, and we both knew that this was a time of change for everyone in America. That change was coming to me especially. After leaving the hospital, I had to limit my physical activity due to being put on blood thinners. I also had to report to the doctor every week to check my iron levels. This meant that I had to scale down my activity for the PT test. This was my first reality; Retirement was becoming a fast option as soon as eligibility permitted. My health and family were the most important thing in my life.

Chapter 16
New Challenges

I was stationed at Ft. Sheridan, Illinois, on January 5, 2009. My husband and I were both excited to be able to live closer to family again. He had family in the Chicago area, and I was closer to Georgia for my family. My first objective, as usual, was to find a church family in the area. I found a warm, loving church and became a member and joined the usher's board. My husband and I both enjoyed living in the Chicago area. Chicago has a great Nigerian culture, great food, and access to the military base.

After settling into the Chicago area, I took my son who was two and a half years old to the pediatrician. He was not meeting some speech and language milestones. The doctor referred him to a specialist who used a form of "play therapy" to try to get my son to communicate. He was eventually diagnosed with Attention Deficit Hyperactivity Disorder (ADHD).

I have always tried to maintain a solid relationship with my son. When he was diagnosed, I realized I needed to be his advocate. I wanted to establish trust with my son like the trust that I had with my mom. I didn't want my son to feel different from other children or isolated. I also wanted him to know that he could express his feelings about anything that was on his mind. I told him if he ever felt like he was having a problem, he could have "immunity."

Meaning, if anything was troubling him, he could tell me about it without the fear of getting into trouble.

I would rather for him to speak about what was going on, rather than feeling like he had to carry that stress alone. He should not have to be forced to speak with someone outside of the family. Many children have their own observations of this world. They process totally different from adults. They can, at times, experience anxiety about feelings and self-image. Our "immunity" was a way to assure my son a process of communicating with me safely.

My son has always been one of the largest or the tallest kids in his peer group. This can be a very positive thing. Also, it has some negative aspects. Physically, it can be an advantage. He has always performed well in sports. One of the problems with being taller or bigger is that the people often assume that he is older than he is. When you are viewed as being older, people can put expectations on you that may be challenging to fulfill. This can be unfair to a child and even stressful.

I learned early that I needed to be an active parent with my son. I wanted to always be present for school meetings. I wanted to be the person who always knew what was being said and was familiar with the terminology and discussions. It was important to make sure he received the right care. As a Soldier, it can be difficult to get the best possible services needed for your child. The military assignments given are often located away from specialists needed for a family. The only choice available for us was to drive our son to a specialist in Milwaukee for his appointments. After he received his diagnosis, we moved directly across the street from the base so that I could spend more time at home.

In July 2010, I received a Certificate of Achievement in Six Sigma Green Belt from Villanova University. This certification was on how "to make better business decisions, and how to apply tools and methodologies to solve critical business process problems" (Villanova University). The following month, I returned to graduate school. I always had a fascination with psychology, so I enrolled once again taking advantage of my benefits. I learned about my son's new diagnosis and ways that I could work with him effectively. I learned about myself and the adjustment that it would take to be successful on my journey leaving the military.

My decision to complete post graduate studies was a pivotal step in my career path. It would change my journey as well as impact my family's future. The fast pace of the military taught me how to grasp knowledge quickly. My post graduate studies taught me how to calm down during stress.

When my son was diagnosed Attention Deficit Hyperactivity Disorder, I mastered the art of redirection. The purpose of this technique was taking his focus off what he was doing and channeling those energies in another direction. I learned that this redirection technique applied to myself. I learned in times of anxiety to channel my mind in another direction through prayer or journaling. I became better at sustaining those long hours of continuous military operations without feeling drained. When I began the adjustment to leave the military, I realized that it was a process that was probably going to take some time but was very important.

I had to realize that I hadn't taken time to process each move, each loss of a loved one or other life events. I tried the process of trying to push back those emotions. It was time to seek assistance. I made my appointments at the Veterans Administration (VA) Hospital to obtain information about the services in my community. All Veterans who leave the military do not receive services from the VA. The services that each person is given depends upon characterization of service and other deciding factors. One would think that all Veterans automatically are taken care of upon leaving the service. Unfortunately, it's simply not the case. I submitted my claim for disability with the Veterans Administration.

I graduated on January 31, 2012, with a Master of Psychology degree. Shortly after, I began the application process for retirement.

Chapter 17
The Retirement

On June 29, 2013, one month after my 49th birthday, I gathered with family and friends at a hotel in Illinois to celebrate my retirement. I received: a certificate of appreciation from President Barak Obama, a flag, and my blue ID card. I was glad to be able to finally settle down with my family in one city that we could call "home". My daughter planned the party with my friends who lived in Minnesota, Tennessee, and Georgia.

I enjoyed looking back at the journey we had overcome. The thoughts of celebrating our future filled my mind. I had been spending time on my terminal leave from the Army, planning out new chapters in our lives. Looking ahead at the plans, my concern with free time, and how to manage flooded my thoughts. I wondered how to occupy the excess of my time. I knew it would be a big adjustment. My family time would become my new focus.

***

The Retirement Speech

"When new Soldiers come into the Army, each are assigned a sponsor that provides tools needed to be successful. A sponsor will provide you with information about whom and where to report. A sponsor will also provide signs to follow that will show the best direction. A good sponsor stands in the gap to keep you out the line of fire.

"I don't have any great and wonderful stories to tell you of anything that I could have done to have been blessed with passing the entrance tests into the military, serving for over 26 years and then retiring.

"Truth be told I was sponsored. The Lord has protected and guided me from the very beginning. He first blessed me with a mom who could not drive a car but who had an incredible faith. Those times we walked for miles to the grocery store; God was preparing me for basic training.

"Another credit to my sponsor is making it possible for me to have never been deployed to war. I know that this accomplishment was His because He truly knows that I can't shoot that well.

"There is one positive thing that I can take credit for in my life is that....I TRUST MY SPONSOR!"

SFC Tammie L. Otukwu, US Army (retired)

CHAPTER 18
THE VETERAN

On August 1, 2013, I officially became a Veteran. I was 49 years old. This was a time of reflection that I had not been able to take in years. I served in the Army for more than half of my life. The military is a very fast-moving train that does not stop. Many Soldiers I have worked with in my career were some of the most dynamic and dedicated individuals I have ever met. They were often called into missions and held jobs that the average person would not have completed.

Soldiers are often asked to multitask, work many hours a day, and sometimes work for weeks or months at a time. This can take a physical and mental toll on any person. Many Soldiers are moving so fast, they can get caught up and neglect themselves. This is because of the mindset of the average military Soldier. They are loyal to the country, their mission, and take great pride to ensure it is carried out to the end. This mindset gives reason to rejoin the ranks if requested. To leave the service and become a Veteran is an honor, yet a great adjustment. It is not easy due to the mental and physical release to turn over the job.

The adjustment that one must make to survive in the service (in my opinion) is very different than the adjustment a person must make when their time in service has ended. Service members often take for granted the preparation needed to one day exit the military and become a Veteran. I had many concerns upon leaving the service and returning to the civilian community. Some of my concerns were: having job security, balancing a budget, navigating the civilian culture, and learning new systems and processes, to name a few. When I left the service and started speaking with other military service members, I realized they too had some of the same fears and concerns.

One of the first things I did after leaving the military was reconnect with family. I went home to Georgia. I helped take care of my Mom. I realized I missed so many life events. It felt like I had been living on another planet.

Family members had grown up, married, had children and moved away. To ease my transition, I connected with family members who had been familiar with the service. I found being able to speak with them about my feelings was very comforting. I found myself speaking a lot about my military career, just like my dad had done in my childhood.

I began to understand why he needed to vent about his events in the service. One other important step was connecting with friends, who had also retired. They were able to assist me with the transition by sharing some of the ways they had been able to be successful.

***

I have always enjoyed volunteering in the community, so I joined a couple of organizations as a way of connecting to people and being of service. I created some accounts via social media to be able to continue to connect with family and friends. My next goal to help my transition to the civilian community was returning to Nashville. I wanted to rejoin my church. I desired to be close to my family and live in a very military friendly city. When I returned to Lake Providence Missionary Baptist Church to rejoin, I found my Pastor and church family still going strong in the community. The church had relocated to a larger sanctuary. The population had grown to well over four thousand members.

During the months before leaving the service, I looked over my resume. I realized it needed translation from my military skills into terminology suited for the civilian world. So much had changed since I updated my resume, including the process of submitting documents. I remember when a hard copy resume would be enough to get an interview.

Now, in the age of social media, I realized that my electronic resume was viewed by many people. Also, I created some professional online profiles to connect with people who were in my career field. I felt confident that my military skills could be valuable to many civilian positions similar to my jobs in the military. I had to take into consideration the civilian work atmosphere being different from the military environment, the language, the chain of command and the structure.

Finances are another area that can be challenging for a veteran because so many services are taken care of on active duty. Service members do not have the concern of medical insurance, life insurance, hospital fees or many of the utilities because the base often has these included. The military even provides for the cost of the uniform. Veterans leaving the service often must purchase an entire new wardrobe depending on the work environment. They might also have to now find a medical doctor, dentist or other providers.

Many Veterans struggle with finances because they get accustomed to a steady paycheck. When you leave the military, you also must leave a mindset and get disciplined and create a budget and a lifestyle that fits your situation. You also must communicate with your family members so that everyone is on the same sheet of music.

Chapter 19
The Case Manager

One of my first positions after retiring from the Army was working as a Supportive Services for Veterans and Families (SSVF) Case Manager for a Non-profit Organization called Operation Stand Down in Nashville. This organization has funding that provides housing, job assistance, counseling and training for Veterans who qualify. I came to the organization as a Veteran looking for assistance with my resume so that I could find a job.

The Deputy Executive Director who worked there at the time saw my resume and offered me the position. In this role, I was able to provide housing to homeless Veterans who qualified for the program. I had no idea that the number of homeless Veterans existed in the U. S. had increased at such a high rate. During my employment, I was also given an opportunity to speak as a panelist for the public screening called "Women in War" which was presented on July 24, 2015. This was a great opportunity to be able to look at a reflection of the history of women in the military.

After the screening, the panel received questions from the audience, and we had an opportunity to provide information about our experiences in the service. It was a very rewarding, challenging and educational experience to be a part of this panel and work for the nonprofit. This position allowed me to really tap into experience and education because the Veterans told stories to which I could relate.

I also met many other Veterans in the Nashville area. The stories of so many homeless Veterans have not been told because some of the individuals have been cast aside by society. I will never forget the many faces and stories of these individuals who came to the nonprofit. Some individuals who had served so proudly struggled to live up to the many expectations of the service. Some of those who had so much experience gained through the service had medical issues and could not maintain employment. The employees that work at this nonprofit were very passionate about helping the many Veterans seeking assistance.

Our Story: Operation Stand Down Tennessee

OSDTN engages, equips and empowers Veterans transitioning from successful military service to civilian life. The organization helps Veterans who have just out-processed, as well as those who completed their service years ago.

Originally focusing on assisting homeless Veterans and homelessness prevention, OSDTN's services have grown to include: Veterans benefits education and access, job readiness and placement, financial counseling, legal assistance, housing, family support, basic needs and service referrals. OSDTN also operates 12th Ave Thrift, a discount store employing Veterans.

Vision: Since 1993 Empowering Veterans to Achieve Their American Dream.

Mission: OSDTN Assists Veterans and their Families so They Can Be Self-Sustaining and Better Connected to the Community.

Value Statement: With Empathy and Expertise, OSDTN Engages, Equips and Empowers Veterans.

FAQ ABOUT HOMELESS VETERANS

Who are homeless veterans?

The U.S. Department of Veterans Affairs (VA) states that the nation's homeless veterans are predominantly male with roughly 9% being female. The majority are single; live in urban areas; and suffer from mental illness, alcohol and/or substance abuse, or co-occurring disorders. About 11% of the adult homeless population are veterans.

Roughly 45% of all homeless veterans are African American or Hispanic, despite only accounting for 10.4% and 3.4% of the U.S. veteran population, respectively.

Homeless veterans are younger on average than the total veteran population. Approximately 9% are between the ages of 18 and 30, and 41% are between the ages of 31 and 50. Conversely, only 5% of all veterans are between the ages of 18 and 30, and less than 23% are between 31 and 50.

America's homeless veterans have served in World War II, the Korean War, Cold War, Vietnam War, Grenada, Panama, Lebanon, Persian Gulf War, Afghanistan and Iraq (OEF/OIF), and the military's anti-drug cultivation efforts in South America. Nearly half of homeless veterans served during the Vietnam era. Two-thirds served our country for at least three years, and one-third were stationed in a war zone.

About 1.4 million other veterans, meanwhile, are considered at risk of homelessness due to poverty, lack of support networks, and dismal living conditions in overcrowded or substandard housing.

CHAPTER 20
THE WAR

I have often asked myself how someone could find themselves in an emotional crisis after leaving the service. There is no simple answer to that question in my opinion. It all depends upon the person and the situation. I personally think anyone can experience hurt and pain with the transition from the military just as if he or she lost a loved one.

If you take a person and put them into a very structured environment, he or she can become so acclimated that their behavior and thought processes start to change. Then, when you take that same person out of the environment and put them into another environment that has vastly different dynamics and structure, it can take a while for the adjustment to take place.

During the time most Veterans are making the adjustment, life continues to go on and they are left to work and continue to take care of themselves and their families. Some of them do not seek assistance in making the adjustment from military to civilian. I think that this mindset is simply part of the military culture.

Another important factor is that the bonds that are formed in the military are often not the same as in the civilian world. As a result, some Veterans isolate. Then you must consider all the trauma that might have occurred within an environment that the person might have experienced. That is when anyone can start to have issues. When I left the military, my passion became providing Veterans with information to assist them on their journey.

I had the opportunity to meet a church member who had extensive experience in working with Veterans. He was a Social Worker with over 30 years of experience and a very positive man of God. He too was a Veteran. His name is Clyde Poag. He introduced some very important concepts through Rational Behavior Therapy, (RBT) "a form of cognitive behavioral therapy developed by psychiatrist Maxie Clarence Maultsby Jr. A professor at the Medical College at Howard University.

RBT is designed to be a short-term therapy which is based on "discovering an unsuspected problem which creates unwanted mental, emotional and physical behaviors." Clyde worked very extensively with Mr. Maultsby. He taught me very important tools to rely on when having to face any of life's adjustments. In my opinion, the hardest struggle experienced was going from the military to civilian life in the mind. He taught me, "it is not what happens to you but what you tell yourself about what happens to you".

I had earned my master's degree in Psychology. I found this training very interesting and helpful. It taught me a new way of thinking that helped me with my transition to the civilian world. If I can share a few things about making the adjustment upon leaving the military, it would be to maintain your faith, cultivate those positive relationships, find resources, and never give up on yourself or "the war".

BIOGRAPHY OF
CLYDE A. POAG, LCSW, ACSW

Clyde Poag is presently President and CEO of Rational Training and Counseling Associates, a counseling, consulting, and training agency founded in Grand Rapids, Michigan, and presently located in Nashville, Tennessee. This Organization was established in 1998 by Clyde and his wife Donna. He is a graduate of Tennessee State University and received his B.S. degree in 1971. He received his Master of Social Work Degree from the St. Louis University School of Social Work in 1975.

Clyde grew up in East St. Louis, Illinois, and is a 1962 graduate of Lincoln Senior High School. He is married and the father of two adult children. They presently have two grandsons. He and his wife Donna have been married 50 years. In 1994 Clyde and Donna were both inducted into the Lincoln High School Hall of Fame.

They are members of Lake Providence Missionary Baptist Church and leaders of the Couples Ministry of the church. He has assisted many churches and families in the Grand Rapids area by providing counseling and training to church members and their families.

Clyde served in the United States Army three years as a Medic and as a Clinical Social Work Psychology Specialist. He returned to East St. Louis in 1971 and was appointed Coordinator of After - Care services for the St. Clair County Community Mental Health Center. In 1976 he was - appointed to a clinical Social Work position with the Department of Veterans Affairs Out Patient Clinic in Grand Rapids.

In 1981 he was appointed Team Leader and established the Grand Rapids Veterans Readjustment Counseling Center. Clyde received 16 Outstanding and Superior Performance awards from the Department of Veterans Affairs and served on numerous committees, including the Bio-Ethics Committee, Homeless Veterans Working Group, and five years as Chairman of the Readjustment Counseling Service African-American Veterans Working Group. He authored the treatment and outreach chapters of the African-American Veterans Working group paper. He worked 22 years as a crisis care clinician with Blodgett Memorial Medical Center in the emergency room.

In 1994, he received a Congressional Award for his work with homeless veterans following Hurricane Andrew and for the establishment of a 3.5-million-dollar program for homeless veterans which was the first Shelter Plus Care Homeless Veterans Program in the country. Also, in 1994 he received the Giants Award from the Grand Rapids Community College for Community Service. He received the "Game Changers Award" from the East St. Louis NAACP. Publications include the treatment of African American Veterans and Outreach Chapters of the African-American Veterans Working Group publication, and the Department of Veterans Affairs Bio-Ethics Report, and an article on fatherless boys.

Following his retirement from the Department of Veterans Affairs he was appointed Assistant Vice-President of Clinical Services, and Clinical Director for Matrix Human Services in Detroit Michigan. He supervised clinical programs of Matrix including "Off the Streets" a program for homeless and runaway youth and The Barrett House: a program for girls adjudicated by the courts.

Former Adjunct Professor of Social Studies at the Grand Rapids Community College Grand Rapids Michigan, Jordan College, and Tennessee State University Nashville Tennessee, He was Clinical Consultant to the Kent County Michigan Juvenile Detention Facility.

He is a certified trainer in the Effective Black Parenting Program and the Healthy Marriage Healthy Relationships programs. Former clients include the Department of the Army, and the U.S. Air Force, The National Medical Association, The Kent County Michigan Chapter of the American Red Cross, Detroit Head Start, and Oakland County Michigan Family Services.

He is a Cognitive Behavior Therapist and provides individual, family and group therapy to families in Middle Tennessee. Presently he and his wife provide training to individuals, groups, community and governmental agencies. He was trained by Dr. Maxie C. Maultsby, Jr. and has conducted workshops with Dr. Maultsby.

He served as consultant to the "Gang Violence reduction project" of Kent County and consultant to the Kent County Juvenile Detention facility. He has been consultant to West Michigan Hispanic Center and Urban Young

Life. He is presently a critical incidence responder for R3 Continuum, an international organization that responds to natural disasters such as hurricane Katrina or man-made disasters such as The Opry Mills shooting and Columbine. He has served as an adjunct Professor to several colleges and universities, including his alma mater Tennessee State University. He served as therapist for a self-help group in Nashville called "Moms over Murder." A group of mothers whose sons were murdered in the Nashville Tennessee.

www.ingramcontent.com/pod-product-compliance
Lightning Source LLC
Chambersburg PA
CBHW072214070526
44585CB00015B/1336